IS THERE ALIEN LIFE IN OUTER SPACE?

DISCOVER THE SCIENCE BEHIND ASTROBIOLOGY

(AS-troh-by-oh-luh-jee)

Written by Olivia Watson

Illustrated by Denis Alonso

Words that are tricky to understand are in **bold**. Find out what they mean in the glossary.

Words that are difficult to say are in *italics*. Find out how to say them at the back of the book.

WHAT IS ASTROBIOLOGY?

Astrobiology is the study of life in the universe. Scientists look at what is needed for life, as well as the possibility of whether there is life in outer space.

The scientists who search for life in outer space are called **ASTROBIOLOGISTS**.

Our universe is made up of billions and billions of stars and planets. When we look up at the night sky, we can only see a tiny number of them. Scientists started to wonder if there were any planets like Earth that were home to living things...

For a long time, scientists called *astronomers* studied the night sky with just their eyes. They had no special tools so couldn't see very much. But then four hundred years ago, **telescopes** were invented.

An Italian man called *Galileo Galilei* made
a very special telescope. He used it to see
the stars and planets in more detail
than ever before. He made lots of
INCREDIBLE DISCOVERIES!

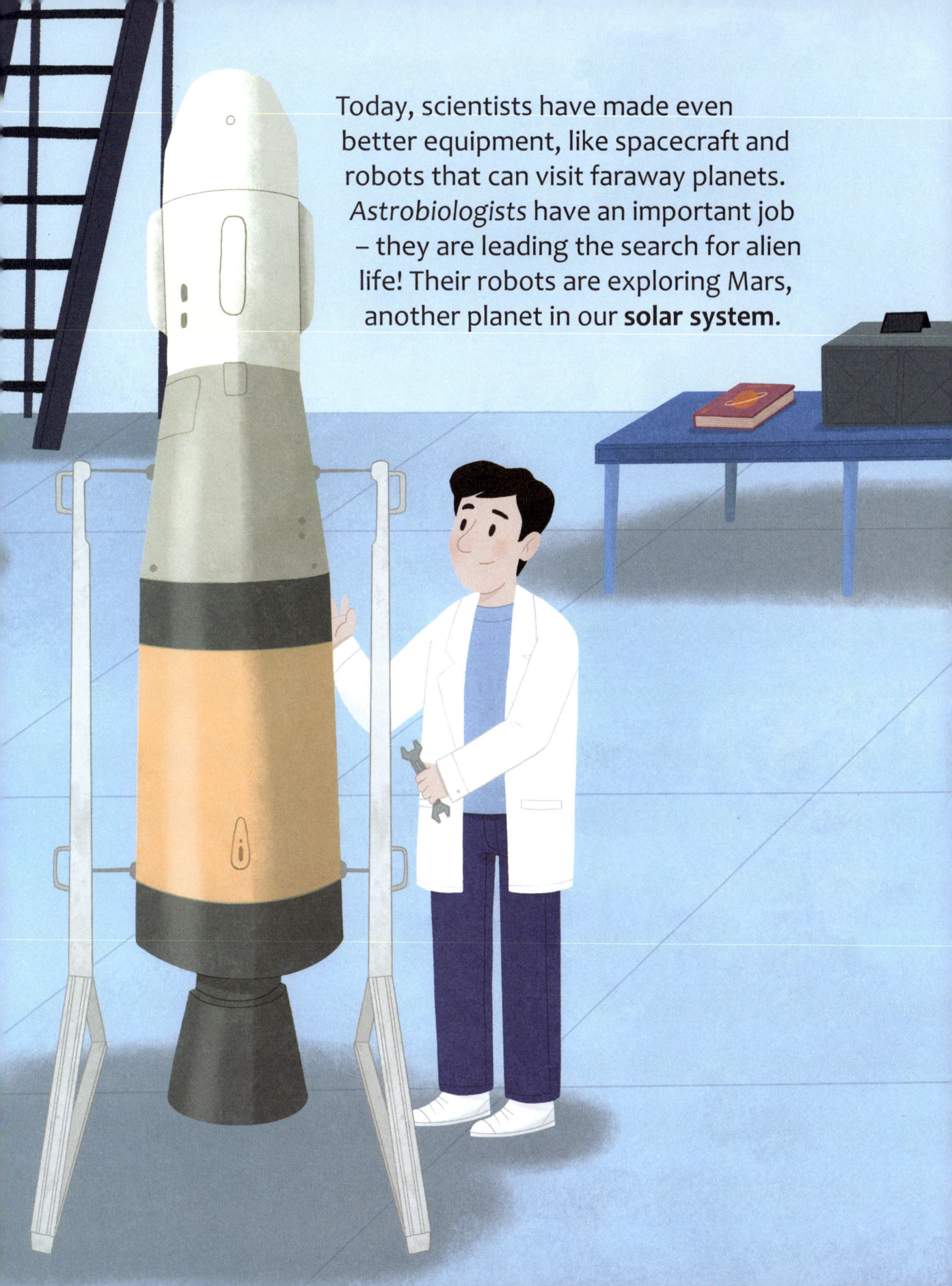

Today, scientists have made even better equipment, like spacecraft and robots that can visit faraway planets. *Astrobiologists* have an important job – they are leading the search for alien life! Their robots are exploring Mars, another planet in our **solar system**.

NASA's robots haven't yet found living things, but they have made some amazing discoveries. Dry, dusty Mars may once have had flowing rivers of water! **Rovers** even found that long ago Mars could have had the right conditions for **microscopic** life, but not today.

To survive on Earth, living things need water, energy, **nutrients**, and a stable **atmosphere** containing **oxygen** to breathe. So, astrobiologists think it's possible that alien life might need one or more of these things to live too.

Our Earth is very special. It's in a perfect place in our solar system called the **"habitable zone"**. If it were closer to the Sun, it would be brown, burnt, and too hot for anything to live. If it were further away, it would be too cold.

The search is on for planets like Earth! But how do we know if distant planets have life? Most planets are too far away for rovers to visit, so astrobiologists use super strong telescopes instead.

They are always finding new planets
DEEP IN OUTER SPACE!

Astrobiologists have discovered more than 5,000 planets outside our solar system! They are called **exoplanets**. Although some of them are in their habitable zone, that doesn't mean the planet is perfect for life...

WASP-12b

SS Cancri e

KOI-55 b

Kepler-7b

GJ 15 A b

HD 189733 b

TOI-3757 b

Proxima Centauri b

TrES-2 b

GJ 504b

AU Microscopii b

Some scientists now think that alien life would be very different from human life. Many animals on Earth have **adapted** to extreme conditions where humans couldn't survive. Maybe aliens could do the same. Could they have...

big wings to fly?

huge and sensitive eyes?

an air bubble to help them breathe?

advanced technology?

Space probes found that it's not just planets that might have life - it's moons too! *Enceladus* is a moon that has an underground ocean! Jets of icy water burst through the surface from the ocean below. The water is rich in **minerals** which are known to help life spread on Earth.

Astrobiologists aren't just searching for signs of life they can see; they're also listening for signs they can hear using **GIANT RADIO TELESCOPES.**

Humans have sent "hello" messages to planets where there might be life. If **intelligent life-forms** exist, they may send messages back to us. We would be able to hear them using these telescopes.

People have reported **UFO** sightings for years. Some say they have seen alien spaceships or found things they can't explain, like weird markings in their crops! They really believe that aliens exist and have visited Earth! But scientists have not yet found any proof.

The incredible size of outer space makes searching for alien life really hard. Astrobiologists haven't found life out there yet. But they're inventing better telescopes and learning new things all the time. So, the search for life in outer space has really only just begun!

ut-of-this world
EQUIPMENT

Astrobiologists work with talented scientists and engineers to invent machines that help them search for life in outer space. Here are some of their most impressive tools.

James Webb TELESCOPE!

This telescope is so strong that it can look back in time to 13.5 billion years ago, when the first stars and galaxies formed! It also helps us see what exoplanets are made of.

Terrific TESS!

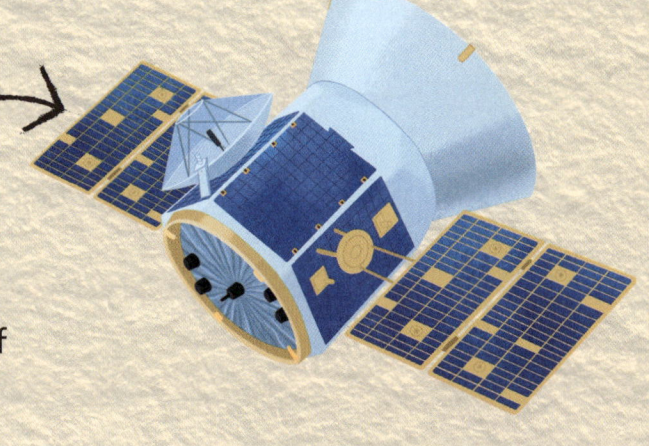

TESS is an exoplanet-finding machine! It has mapped almost all of the sky, finding hundreds of exoplanets, as well as **asteroids**, stars, and galaxies.

Voyager SPACECRAFT!

These machines have flown deeper into outer space than anything before them. They have visited the furthest planets in our solar system and have even entered **interstellar space!**

LANDERS *and* ROVERS!

Many of these robots are currently on Mars! They are controlled by scientists on Earth, who use them to see and test things on Mars.

Kepler Space TELESCOPE!

The Kepler space telescope observed 150,000 stars continuously! It discovered more than 2,600 exoplanets before it was retired.

Amazing
SPACE FACTS

There's so much to discover about the world of astrobiology. Do you know the answers to some of the world's biggest questions about outer space?

WHEN DID HUMANS FIRST JOURNEY INTO SPACE?

In the 1960s! So far, the only place people have walked (except Earth) is the Moon, but scientists are now working on sending humans to Mars!

ARE ASTROBIOLOGISTS THE SAME AS ASTRONOMERS?

Not quite! Astronomers study everything in the universe beyond Earth, whereas astrobiologists focus on whether life exists beyond Earth.

HOW BIG IS THE UNIVERSE?

No one knows for sure
– we can't see it all!
Scientists can measure the
"observable universe",
but the entire universe is
probably much bigger.

WHEN WERE THE FIRST EXOPLANETS DISCOVERED?

In the 1990s! Since then, scientists have found more than
5,000! But they think there may be many more exoplanets
out there, waiting to be discovered...

HOW DO PLANETS GET THEIR NAMES?

The planets in our solar system
were named after Roman gods.
Today, exoplanets are given names
like GJ 504b which tell us how and
where they were found.

GLOSSARY

Adapted – when a living thing has developed special features or skills to help it survive in its environment.

Asteroids – small space rocks that orbit the Sun.

Atmosphere – the gases that surround a planet.

Exoplanets – planets that are outside our solar system.

Habitable zone – the distance a planet needs to be from its star to have running water on its surface.

Intelligent life-forms – living things that can think, learn, and understand things.

Interstellar space – outer space just past the edge of our solar system.

Microscopic – something that is so small it can only be seen with a microscope.

Minerals – substances that are naturally found in things like rocks, sand, and soil.

NASA – the National Aeronautics and Space Administration is an agency that deals with space research and exploration. It's based in the USA.

Nutrients – substances or ingredients that plants and animals need to live and grow.

Observable universe – the parts of the universe that humans are able to see.

Oxygen – an invisible gas in the air that plants produce, and people and animals need to breathe.

Rovers – remote-controlled robots that are built to explore extra-terrestrial planets and moons.

Solar system – the Sun and everything that moves around it.

Space probes – unmanned spacecraft that are sent into space to do research and gather information.

Telescopes – instruments that make faraway objects appear bigger.

UFO – unidentified flying object.

HOW DO I SAY?

Astrobiologists
AS-troh-by-oh-luh-jists

Astrobiology
AS-troh-by-oh-luh-jee

Astronomers
uh-STROH-nuh-mers

Enceladus
en-SELL-uh-dus

Galileo Galilei
gah-li-LAY-oh
gah-li-LAY

THE BIG QUESTIONS ANSWERED

This is more than just a series of books; it is a complete resource.
Accompanying each book is a variety of FREE material to engage curious kids with science.

www.thebigquestionsanswered.com

Use the QR code to visit the website, download free resources, and discover other books in the series.

On the website, find out incredible things about astrobiologists, including what they do, some of their greatest discoveries, and what it takes to become an expert in this field of science.

The material is also available for home or classroom use, supporting all the information in this book.

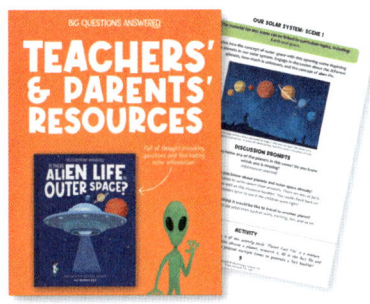

Teachers' & Parents' Resources
With discussion prompts and questions, extra information, and facts around key topics.

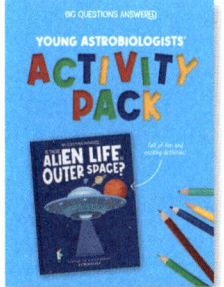

Young Astrobiologists' Activity Pack
Fun activities for wannabe space explorers, including creative writing, drawing, word searches, and much, much more.

BEETLE BOOKS

Beetle Books is an imprint of Hungry Tomato Ltd.

First published in 2024 by Hungry Tomato Ltd
F15, Old Bakery Studios, Blewetts Wharf, Malpas Road,
Truro, Cornwall, TR1 1QH, UK.

ISBN 9781835691298

Copyright © 2024 Hungry Tomato Ltd

With thanks to:
Editor: Holly Thornton
Editor: Millie Burdett
Senior Designer: Amy Harvey
Tim Cook for his valued contribution
The team at Beehive Illustration

Printed and bound in China.

Picture Credits:
(t = top, b = bottom, m = middle, l = left, r = right)
Image courtesy of NASA, public domain: 33bl. Shutterstock: Antares_
StarExplorer 35tl; Dima Zel 32ml; EWY Media 35bl; Frame Stock Footage 34bl;
Nazarii_Neshcherenskyi 35mr; Vadim Sadovski 34mr; Wow Galaxy 33tl